Hiding Away

Anne O'Daly

Brown BEAR BOOKS

Published by Brown Bear Books Ltd

4877 N. Circulo Bujia
Tucson, AZ 85718
USA

and

G14, Regent Studios
1 Thane Villas
London N7 7PH
UK

ISBN 978-1-83572-025-7 (ALB)
ISBN 978-1-83572-031-8 (paperback)
ISBN 978-1-83572-037-0 (ebook)

Library of Congress Cataloging-in-Publication Data available on request

Design Manager: Keith Davis
Children's Publisher: Anne O'Daly
Picture Manager: Sophie Mortimer

Picture Credits
Cover: Shutterstock: Oleksa Antonvo. Interior: iStock: Wirestock 23; Shutterstock: Kurit Afshen 12–13, 22tr,
Sourabh Bharti 18–19, Mark Brandon 10–11, clearviewstock 1, Ethan Daniels 8–9, Divelvanov 22tl, Thierry
Eidenweil 5tl, 14–15, Gerald Robert Fischer 14, FourOaks 4b, Alexandra Giese 20–21, Eugen Haag 21, Stephan
Kerkhofs 9, Anuradha Marwah 18, Rob McGillivray 17, Martin Mecnarowski 5c, OceanEloy 22br, OhmAI2T 11,
Martin Pelanek 22bl, Joana Perchaluk 6, Angela N Perryman 5tr, Photcechcz 4–5b, Kambiz Pourghanad 4–5c,
Francesco Riccalardi 4t, SonjaM 5b, Andrei Stepanov 6–7, Lisdiyanto Suhardjo 12, TETSUSnowdrop 16–17.
t=top, b=bottom, l=left, r=right, c=center
All artwork and other photography Brown Bear Books.

Brown Bear Books has made every attempt to contact the copyright holder.
If you have any information about omissions, please contact: licensing@brownbearbooks.co.uk

Manufactured in the United States of America
CPSIA compliance information: Batch#AG/5663

Websites
The website addresses in this book were valid at the time of going to press. However, it is possible
that contents or addresses may change following publication of this book. No responsibility for
any such changes can be accepted by the author or the publisher. Readers should be supervised
when they access the Internet.

Contents

Meet the Hiders

Animals have clever ways to hide away. Some blend into their surroundings. This is camouflage. Others copy plants or other animals. Meet some amazing hiders!

Hiding Away
Decorator Crab
Penguin
Chameleon
This chameleon's green color matches its background.
Stick Insect

Arctic Fox

This fox lives in the Arctic. There is snow in winter. The fox has a white coat to blend in. In summer, the snow melts. The land is brown. The fox changes color to match.

Name: Arctic fox

Where it lives: the Arctic

What it eats: small mammals, birds, and insects

Type of animal: mammal

Size: 18 to 27 inches (46 to 68 cm)

Body: reddish brown fur that turns white in winter

White fur makes it hard to spot the fox in the snow

The fox turns brown in the summer.

MINI FACTS

The fox has thick fur to keep warm. It has furry paws, too.

Mimic Octopus

A mimic octopus can change shape. It can look like another animal. The octopus copies animals that are dangerous. Predators stay away!

FACT FILE

Name: mimic octopus

Where it lives: in the seas around Indonesia and in the Indian Ocean

What it eats: small fish, worms, and crabs

Type of animal: mollusc

Size: 2 feet (60 cm)

Body: brown and white stripes; eight arms

This octopus is copying a poisonous fish

Mimic octopuses can copy more than 15 animals

MINI FACTS

The octopus has another way to defend itself. It sprays ink at a predator. The ink makes the water cloudy. The octopus can escape.

The octopus pulls its arms close to its body

This octopus is hiding in sand.

Stick Insect

Some animals look like plants.
Stick insects look like twigs or leaves.
They hide from hungry predators.

FACT FILE

Name: stick insect (there are more than 3,500 different kinds)

Where it lives: everywhere in the world except Antarctica

What it eats: leaves and parts of plants

Type of animal: insect

Size: 1 to 25 inches (2.5 to 64 cm)

Body: different stick insects look like sticks, twigs, or leaves

Long, thin legs

MINI FACTS

Some stick insects make a smelly liquid. They spray it if predators get too close.

This butterfly looks like a dead leaf. This helps it hide from predators.

Chameleon

Chameleons usually blend into their background. They can change color. Sometimes they do this to find a mate.

FACT FILE

Name: chameleon

Where it lives: Africa, Asia, and Europe

What it eats: insects

Type of animal: reptile

Size: 2 to 13 inches (5 to 33 cm)

Body: large head and a long tail that can wrap around branches

Bright colors attract a mate

A chameleon has a long, sticky tongue. The tongue is twice as long as the animal's body.

MINI FACTS

Chameleons are darker in the mornings. Dark colors soak up sunshine. This helps the animals warm up.

Decorator Crab

This crab has hooked hairs on its shell.
It picks up seaweed and rocks. It fixes
them to the shell. That helps it blend in.

FACT FILE

Name: decorator crab

Where it lives: oceans around the world

What it eats: seaweed and small fish

Type of animal: arthropod

Size: 0.5 to 9 inches (1 to 23 cm)

Body: tough shell and ten legs

This crab has an animal called a sea anemone on its shell

It is hard to see this decorator crab.

MINI FACTS

Crabs molt when their shell gets too small. They grow a new shell. They sometimes take something useful from the old shell.

Sea anemone

Crab

Penguin

Penguins live near the South Pole.
They hunt for fish in the freezing ocean.
Their colors hide them from predators
and prey.

FACT FILE

Name: emperor penguin

Where it lives: Antarctica

What it eats: fish and squid

Type of animal: bird

Size: 45 inches (115 cm)

Body: black and white feathers with yellow patches on the neck

MINI FACTS

Penguins eat a lot of fish.
They poop every 20 minutes!

Hiding Away

Black back.
It looks like dark
water from above.

White belly. From
below, it looks like
the sun shining
through the water.

Baby penguins are
covered with fluffy
down. It keeps
them warm.

Tiger

Tigers are fierce hunters. They sneak up on prey. Most tigers live in forests and grasslands. Their stripes hide them from other animals.

FACT FILE

Name: Bengal tiger

Where it lives: India and nearby countries

What it eats: mammals such as deer and wild pigs

Type of animal: mammal

Size: 4.5 to 9.2 feet (1.4 to 2.8 m)

Body: strong body with orange and black stripes

Striped fur helps the tiger hide in long grass

Cubs stay with their mom for about two years.

MINI FACTS

Tigers have powerful legs. They can jump forward around 25 feet (7.6 m).

Tigers get close to prey and then pounce

Zebra

Zebras have black and white stripes.
These animals live in big groups.
The zebras stand together. Their stripes mix.
It's hard for a predator to pick out one zebra.

FACT FILE

Name: zebra

Where it lives: grasslands and woodlands in east and south Africa

What it eats: grass and sometimes leaves and tree bark

Type of animal: mammal

Size: 4 to 5 feet (1.2 to 1.5 m) tall at the shoulder

Body: thick body, long head, tail with a tuft

Each zebra has its own pattern

MINI FACTS

A group of zebras is called a dazzle.

Zebras spend up to 18 hours a day eating.

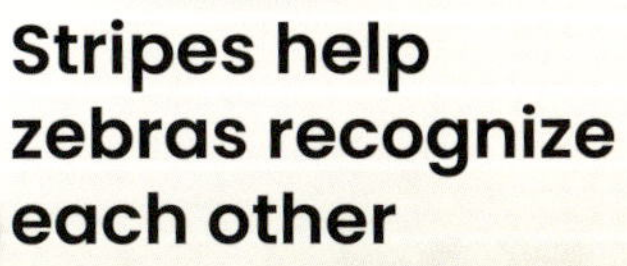

Stripes help zebras recognize each other

The stripes confuse predators

Quiz

Test your skills!
Can you answer these questions?
The answers are on page 24.

1 Why do octopuses spray ink?

2 How can chameleons change color?

3 How far can a tiger jump?

4 What is the name for a group of zebras?

Useful Words

camouflage How an animal hides, usually by blending in with its surroundings.

cells The tiny building blocks that make up living things.

down The small, fluffy feathers that baby birds have.

flippers Body parts that help animals swim.

molt When an animal loses an outer covering and grows a new one.

poisonous Something that can harm or kill an animal.

predator An animal that hunts other animals for food.

prey An animal that is eaten by other animals.

seaweed A plant-like living thing that grows in the sea.

Find Out More

Books

Animal Camouflage in the Ocean, Ruth Owen (Ruby Tuesday Books, 2024)

Animals Hidden in the Forest, Jessica Rusick (Pebble, 2022)

Masters of Camouflage, Nancy Dickmann Brown Bear Books, 2021)

Websites

kids.nationalgeographic.com/wacky-weekend/article/hidden-animals

www.bbc.co.uk/bitesize/articles/zh9frmn

www.ducksters.com/animals/stick_bug.php

Index

Quiz Answers: 1. Octopuses spray ink to get away from predators; **2.** Special cells in the skin change color; **3.** A tiger can jump about 25 feet (7.6 m); **4.** A group of zebras is called a dazzle.